D1736504

TROPICAL RAIN FOREST

BIOMES

Lynn M. Stone

The Rourke Corporation, Inc.
Vero Beach, Florida 32964

PHOTO CREDITS
All photos © Lynn M. Stone

Library of Congress Cataloging-in-Publication Data

Stone, Lynn M.
Tropical rain forest / by Lynn M. Stone.
p. cm. — (Biomes)
Includes index.
Summary: Explores the mysterious green jungles of the tropics which receive one hundred or more inches of rain per year and which are home to a tremendous variety of plants and animals.
ISBN 0-86593-424-X
1. Rain forest ecology—Juvenile literature. 2. Rain forests—Juvenile literature. [1. Rain forest ecology. 2. Rain forests. 3. Ecology.] I. Title. II. Series: Stone, Lynn M. Biomes.
QH541.5R27S765 1996
574.5'2642'0913—dc20 95-46175
CIP
AC

Printed in the USA

TABLE OF CONTENTS

TROPICAL RAIN FOREST

Tropical rain forests are the mysterious green jungles of the tropics. The tropics are hot, humid regions near the Earth's **equator** (ee KWAYT er). The equator is an imaginary line, east to west, around the Earth's middle.

Rain forests are very wet places. They receive 100 or more inches of rain per year. Deserts only have six or eight inches of rain a year.

Many different kinds of plants and animals live in the wet tropical rain forests.

Rain forests in the tropics grow in deep river valleys and high on steep hillsides

KINDS OF RAIN FOREST

Tropical rain forests around the world look much alike. They have layers of tall trees, shorter trees, and shrubs — but tropical rain forests are really not all the same. A rain forest in Central America has different **species** (SPEE sheez), or kinds, of plants and animals than a rain forest in Australia or Southeast Asia.

The rainy coastal woodlands of the Pacific Northwest — Washington, Alaska and British Columbia, Canada — are **temperate** (TEM per it) rain forests. They are not at all like tropical rain forests.

The plants and animals of this tropical rain forest in Central America are different from those in Australia and Southeast Asia

LIFE IN THE RAIN FOREST

The tropical rain forest hums with life. Much of the forest's animal activity, though, is hidden. It happens under leaves, in the soil, and in the high, leafy branches. Sound, rather than sight, often reveals animals. Monkeys and macaws shriek. Jaguars roar and lizards rustle.

The rain forest plants support the animals. Animals eat plants, or they eat the plant-eating animals.

The plants themselves grow by changing sunlight and soil into food.

This green tree python lives in the rain forests of Australia, while the look-alike emerald tree boa lives in South and Central America

RAIN FOREST PLANTS

The largest plants of the tropical rain forest are trees. The tallest trees stand about 150 feet high. They tower over the plant layers below. The tall trees form a canopy, or roof, of the forest.

Along with trees, a tropical rain forest has mushrooms, flowers, shrubs, vines, and unusual plants called **bromeliads** (bro MI lee adz).

Instead of growing in the ground, bromeliads attach to tree bark. Bromeliad "cups" catch rainwater and become homes for insects and climbing frogs.

A poison-arrow frog perches in the leaves of a bromeliad in Costa Rica

The cutting and burning of tropical rain forests threaten many species of plants and animals, some of them not yet discovered!

A scarlet macaw brightens South American rain forests with its fiery feathers

BIRDS OF THE RAIN FOREST

The rain forests are homes for hundreds of species of birds. They eat seeds and fruits, insects, and other small animals.

Some of the birds add splashes of color to the dark, green jungles. The scarlet macaw, for example, is a big, bright red parrot of Central and South America. The toucan's color is in its bright, banana-shaped bill!

Birds in the forest canopy are hard to see, but their loud early-morning chatter is easy to hear.

The keel-billed toucan, a fruit-eater, is called "banana bill" by Central Americans

MAMMALS OF THE RAIN FOREST

A few species of mammals live in the canopy. The noisiest ones are the monkeys.

Several other mammals live below the canopy. They include peccaries, tapirs, and **predators** (PRED uh torz) like the jaguar of South America and the tiger of Asia.

Bats of many kinds live in rain forests. Most eat insects, but some live on fruit and flowers.

The jaguar is a fine climber, and it often rests on tree limbs near South American rivers

OTHER RAIN FOREST ANIMALS

Some of the most important animals in the rain forests are the smallest. Small animals help break down leaves and dead animals into tiny bits. These bits of plants and animals can be used for food by many other living things.

One of the most interesting animals of the rain forests is the poison-arrow frog. It wears bright colors, perhaps to warn predators away from its poisonous flesh.

Several kinds of snakes glide through rain forests. A few, like the eyelash viper of Central America, are poisonous.

An eyelash viper, small but deadly, glides through a Central American rain forest

VISITING THE RAIN FORESTS

Tropical rain forests are true jungles — leafy, dark, humid, and often shiny with raindrops. Tree roots rise like spider legs from the soil. Vines climb tree trunks and dangle from limbs. Lines of leaf-cutter ants march like soldiers across the forest floor.

The forest is alive, but it can be strangely quiet, especially at night.

Few trails lead into the thickest rain forests. Visitors often travel by river to go deep into the forests.

Leaf-cutter ants haul their trimmings down a tree trunk

PROTECTING RAIN FORESTS

Rain forests are disappearing quickly. People in tropical countries cut rain forest trees for homes, furniture, and firewood.

Rain forests are also being cleared, bit by bit, to make space for farming. When forest is cut away, farmers can plant crops in the ground or make pastures for cattle.

Some countries, such as Costa Rica and Belize, are protecting part of their rain forest for wildlife and plants.

Glossary

bromeliad (bro MI lee ad) — a group of rootless, long-leaved plants that grow on bark and branches

equator (ee KWAYT er) — the imaginary line drawn on maps around the Earth's middle at equal distances from the north and south poles

predator (PRED uh tor) — an animal that kills other animals for food

species (SPEE sheez) — within a group of closely-related plants or animals, one certain kind, such as an *eyelash* viper

temperate (TEM per it) — places where temperatures are rarely very hot or very cold

INDEX